AF730870

FUTURE-PROOF

What Every Person and Parent Must Do
Before AI Does It For You

Chauntelle Ricks
Certified AI Consultant

Future-Proof

What Every Person and Parent Must Do Before AI Does It For You

Copyright © 2026 by Chauntelle Ricks

All rights reserved.

No part of this publication may be reproduced, distributed, or transmitted in any form
or by any means, including photocopying, recording, or other electronic or mechanical
methods, without the prior written permission of the publisher, except in the case of
brief quotations embodied in critical reviews and certain other noncommercial uses permitted by copyright law.

Published in the United States of America

Self-Published, 2026

ISBN: 978-1-962942-11-9 (Paperback)

ISBN: 978-1-962942-12-6 (eBook)

Library of Congress Control Number: Applied For

Disclaimer: This book is intended for educational and informational purposes only.
The author has made every effort to ensure accuracy of information at time of publication.
Statistical data and research findings are cited to the best of the author's knowledge.
Readers should consult qualified professionals for career, legal, or financial decisions.

Cover design by Chauntelle Ricks

Interior design formatted for Amazon KDP

https://stan.store/edgeeffectai

First Edition

Printed in the United States of America

CONTENTS

Introduction A Note Before We Begin

PART ONE Wake Up

Chapter 1 *The House Is Already On Fire*

Chapter 2 *Who Wins. And It's Not Who You Think.*

PART TWO The Adult Playbook

Chapter 3 *You Don't Need to Code. You Need to Think.*

Chapter 4 *Your Career Audit. A Blunt Conversation.*

Chapter 5 *Robot-Proof. The Skills That Actually Last.*

PART THREE The Parent Playbook

Chapter 6 *The School System Wasn't Built for This World.*

Chapter 7 *What to Stop Optimizing Your Kid For.*

Chapter 8 *What to Start Doing At Home Right Now.*

PART FOUR The Moral Frontier

Chapter 9 *The Church Can't Afford to Sit This One Out.*

PART FIVE The Decision

Chapter 10 *The Only Real Risk Is Standing Still.*

References Sources and Citations

A NOTE BEFORE WE BEGIN

This is a short book by design. Not because the subject is simple, but because simplicity is a form of respect. You are busy. The world is moving fast. And the most important ideas do not require hundreds of pages to land with force.

Each chapter is built around one clear idea, supported by real data, grounded in practical application, and closed with prompts and actions you can use immediately. The prompts are not optional extras. They are the point. Reading about the AI transition is interesting. Doing something about it is what changes your trajectory.

A note on the voice of this book: I write from experience, both my own and from the hundreds of conversations I have had with professionals, parents, educators, and faith leaders who are navigating this shift in real time. Where I share a personal moment, it is because the best way to make an idea land is to make it human first. Where I cite data, it is because credible evidence matters, and you deserve to know the difference between an opinion and a finding.

One more thing. This book is deliberately free of em dashes, jargon, and the kind of breathless tech-industry enthusiasm that makes most AI books exhausting to read. What you will find instead is straight talk, honest assessment, and a deep conviction that the people who navigate this transition well will not be the most technically gifted. They will be the most willing to engage.

Let's begin.

PART ONE

Wake Up

CHAPTER 1

The House Is Already On Fire

Why the urgency is real, and why most people are still asleep

The first time I asked an AI to do something I thought only a human could do, I sat back in my chair and stared at the screen for a long moment. Not because it failed. Because it nailed it. Completely. In about eleven seconds.

I had spent the better part of a morning preparing a framework for a training session, something that typically took me two or three hours to pull together. I typed a prompt, gave some context, and watched it build what I had been building in my head for weeks. And in that moment, I had a very honest thought: how many people are walking into work tomorrow still believing that their value lies in the thing I just watched a machine do in under a minute?

That question is the reason this book exists.

> ***This isn't a book about artificial intelligence. It's a book about you, and whether you're ready for what's already happening.***

STOP TREATING THIS LIKE TOMORROW'S PROBLEM

Here's the thing about major disruptions: they rarely arrive with a warning label. The factory workers who lost their jobs to automation in the 1980s didn't get a five-year notice. The travel agents, video rental clerks, and newspaper editors who watched the internet gut their industries in the 2000s didn't see a countdown clock. And the millions of people whose roles are being reshaped by AI right now are, in many cases, still not paying attention.

AI isn't a future problem. It is a right-now reality. By 2024, AI was already handling legal research at major law firms, writing first drafts of marketing copy for Fortune 500 companies, screening medical images for early signs of cancer, managing customer service interactions for millions of businesses, and grading student essays at universities. This is not science fiction. This is Monday morning.

And the pace of change isn't slowing down. It's accelerating. The difference between what AI could do in 2022 and what it can do today is not small. It's a generational leap compressed into twenty-four months. The trajectory we're on means that whatever AI can't do today, it will very likely be able to do within three to five years.

39% *of existing skills will be transformed or obsolete by 2030. World Economic Forum, Future of Jobs Report, 2025.1*

92 Million *jobs displaced by 2030, offset by 170 million new roles, a net gain of 78 million. World Economic Forum, 2025.1*

The math looks fine on paper, more jobs created than lost, a net gain of 78 million by 2030. But here's the problem with those numbers that nobody talks about: the people losing those 92 million jobs aren't automatically qualified for the 170 million new ones. That gap, between the work that's disappearing and the work that's emerging, is exactly where people get left behind. And it's not a gap that fills itself.

THE COMFORTABLE LIE WE TELL OURSELVES

There's a version of this conversation that's been happening for years, something like this: AI will take the boring, repetitive jobs, but the creative and complex work, the human work, will always need people. And to be fair, that's partially true. But it's also the kind of reassurance that keeps people from taking action, and that's where it becomes dangerous.

AI is no longer just automating data entry and assembly lines. It is writing novels, composing music, generating architectural blueprints, passing the bar exam, scoring in the top percentile on medical licensing tests, and producing financial analyses that rival those of trained analysts. The line between boring repetitive work and skilled professional work has moved considerably, and it is still moving.

The comfortable lie is that this is happening to other people, in other industries, in other countries. The reality is that every sector, healthcare, education, law, finance, ministry, real estate, marketing, manufacturing, retail, is being reshaped right now. And the people who will come out ahead aren't necessarily the smartest or the most educated. They're the ones who stopped waiting to see what would happen and started moving.

> ***The people who get left behind in every major technological shift aren't the least capable. They're the least willing to move.***

WHAT THIS BOOK IS, AND ISN'T

This is not a doom-and-gloom book. The future, even with AI in it, especially with AI in it, holds more opportunity than most people can currently imagine. The goal of this book is to make sure you're one of the people positioned to capture that opportunity rather than one of the people blindsided by the shift.

This book is also not a tech manual. You will not find tutorials here for how to build AI models or write code. What you will find is a clear, practical, no-nonsense guide to what every person, regardless of their technical background, needs to understand, develop, and do to thrive in a world where AI is as common as a smartphone.

And for the parents reading this: there is an entire section of this book dedicated specifically to you. Because the decisions you make in the next two to three years about how you educate, prepare, and equip your children will matter more than almost any other parenting decision you will make. The school system isn't keeping up. That means you have to.

We'll also talk about something that almost no one in the AI conversation is talking about: the moral dimension. The role that faith communities, in particular, need to play in making sure that, as AI becomes more powerful, it is guided by something more lasting than profit. But we'll get there. First, let's make sure you fully understand what you're dealing with.

THE SPEED NOBODY IS READY FOR

One of the most disorienting things about the AI era is the speed at which it's moving. Most major technological shifts have given society a decade or two to adjust. The printing press took a century to fully reshape European society. The Industrial Revolution unfolded over multiple generations. Even the internet, which transformed virtually every aspect of modern life, gave people roughly twenty years to adapt their careers, businesses, and institutions.

AI is not moving on that timeline. The leap from GPT-3 to GPT-4, two AI systems released less than two years apart, represented a capability jump that experts described as stunning even to those building it. The capabilities that felt like science fiction in 2020 are workplace tools in 2026. The capabilities that seem advanced today will be standard features in three years.

This isn't meant to paralyze you. It's meant to calibrate your sense of urgency. Because the one thing that is absolutely clear from every economist, technologist, and workforce researcher studying this shift is the same: the people who start adapting now, who build the right skills, the right mindsets, and the right habits today, will have an enormous advantage over the people who wait until the disruption is impossible to ignore.

The house is on fire. You can see the smoke. The only real question is what you're going to do about it.

TRY THIS RIGHT NOW

Personal Reflection

What parts of my current work or daily life could be done at least partially by an AI tool today? And how does that make me feel?

AI Conversation Starter

I work in [your industry] as a [your role]. Be honest with me: which parts of my job are most vulnerable to AI automation in the next three years, and what should I be doing about it right now?

Family Conversation

Ask your kids tonight at dinner: if a robot could do your homework, what would you spend that time learning instead? Then really listen to the answer.

DO THIS NOW

1. Write down the five most repetitive tasks in your work or daily life. These are the first things AI will replace, and knowing them gives you a head start on deciding what to do next.

2. Download one AI tool this week: ChatGPT, Claude, Gemini, or Copilot, and use it on a real task, not just to test it. Give it something that actually matters and see what it does.

3. Have an honest ten-minute conversation with someone you trust about how prepared you both actually feel. Not how prepared you think you should say you are. How prepared you actually are.

NEXT: Chapter 2 — Who Wins. And It's Not Who You Think.

CHAPTER 2

Who Wins. And It's Not Who You Think.

Why the AI era rewards adaptability over credentials, and curiosity over comfort

---

I have watched incredibly smart people freeze. People with advanced degrees, years of experience, and genuine talent who sat across from a new challenge and simply stopped. Not because they lacked the ability to figure it out. Because they had spent so long being the expert in the room, the idea of being a beginner again felt unbearable.

And I have watched people with far fewer credentials, far less experience, and far simpler resumes run circles around them. Not because they were smarter. Because they were willing. Willing to try something new, look foolish for a minute, ask the basic question, and move before they had all the answers.

That gap, between the willing and the unwilling, is going to define winners and losers in the AI era more than any other single factor. And the sooner you understand why, the better positioned you will be to end up on the right side of it.

> ***The people who get left behind in every major technological shift aren't the least capable. They're the least willing to move.***

THE CREDENTIAL TRAP

For most of the twentieth century, the path to a good life was straightforward: get educated, get credentialed, get hired, stay put. The system rewarded people who could learn a defined body of knowledge, demonstrate mastery of it, and apply it reliably over a long career. Stability was the prize. Predictability was the point.

That system worked because the world changed slowly enough for credentials to hold their value. A degree earned in 1975 was still largely relevant in 1995. A professional certification from 2005 still carried weight in 2015. The shelf life of knowledge was long, and the institutions that conferred it could be trusted to keep up with what the market needed.

That era is over. Not because education stopped mattering, but because the pace of change now outstrips any institution's ability to credential people fast enough. The World Economic Forum's 2025 Future of Jobs Report found that 39% of existing workforce skills will be transformed or obsolete by 2030.1 That is not a slow drift. That is a fundamental restructuring of what knowledge is worth.

The credential trap is believing that what got you here will get you there. It won't. Not because your past learning has no value, but because the half-life of specific technical knowledge is shrinking fast, and the people who keep winning are the ones who treat learning as a permanent practice rather than a completed phase of their life.

59 out of 100 *workers will need reskilling by 2030, and 11 of them won't receive it. World Economic Forum, Future of Jobs Report, 2025.1*

WHAT THE WINNERS ACTUALLY HAVE IN COMMON

McKinsey, the World Economic Forum, Harvard Business Review, and virtually every major research organization studying the future of work have spent the last several years trying to answer the same question: what separates the people who thrive in periods of rapid technological change from those who struggle?

The answer is not a specific technical skill set. It is not a particular degree. It is not even industry-specific knowledge, though it helps. The research consistently points to a cluster of human capabilities that cut across every profession, sector, and educational level: critical thinking, the ability to convey complex ideas clearly, creativity, and adaptability. The willingness to update your understanding when new information arrives, to try approaches that might not work, and to keep moving when the path isn't clear.

None of these are new ideas. What is new is how much more valuable they have become now that AI can handle the routine, the repetitive, and the purely analytical. The more AI absorbs those functions, the more the uniquely human capabilities become the differentiator. The floor is rising fast for everyone. The ceiling for people who have developed these skills is rising faster.

> ***The more AI absorbs the routine and the repetitive, the more your human capabilities become your greatest competitive advantage.***

THE CURIOSITY ADVANTAGE

Here is a finding that should reframe how you think about your own development and your children's: the research on who adapts well to major change consistently points not to intelligence, not to education level, and not to age. It points to curiosity.

Curious people ask questions that others assume are already answered. They read outside their field. They experiment with tools before they're told to. They find the adjacent idea interesting rather than threatening. And in a period when the tools, processes, and required skills are shifting faster than any institution can track, the person who stays curious has a structural advantage over the person who stops exploring when they get comfortable.

A 2023 study from the London School of Economics found that intellectual curiosity was one of the strongest predictors of career resilience across industries facing automation.2 Not IQ. Not years of experience. Curiosity. The willingness to keep asking what else is possible.

Curiosity *ranked among the top predictors of career resilience across industries facing automation. London School of Economics, 2023.2*

AGE IS NOT THE EXCUSE YOU THINK IT IS

One of the most common things people say when they hear about AI and the need to adapt is some version of: I am too old for this. And every time I hear it, I want to push back hard. Not because age doesn't bring real challenges when learning new things. It does. But the research on adult learning and career reinvention tells a very different story than most people assume.

A 2022 AARP study found that workers over 50 who engaged in regular upskilling were not only able to keep pace with younger colleagues in AI-adjacent roles, but in many cases outperformed them in the areas that matter most: judgment, communication, and the ability to apply new tools to complex real-world problems.3

Experience is not the liability it's often framed as. It's a foundation, if you're willing to keep building on it.

The people who genuinely struggle with adaptation aren't struggling because of their age. They are struggling because of their mindset. The fixed belief that their identity is tied to what they already know makes learning something new feel like a threat rather than an opportunity. And that is a belief system, not a biological limit. It can be changed.

If you are reading this and feeling the quiet anxiety of not knowing where to start, that is completely normal. But anxiety and willingness can coexist. You do not have to feel ready to begin. You just have to begin.

THE SOFT SKILLS THAT ARE NOW HARD REQUIREMENTS

For decades, skills like emotional intelligence, empathy, active listening, and ethical judgment were described as soft skills, nice to have, the kind of thing that got a polite mention in a performance review but rarely drove a hiring decision or a promotion. That framing is changing rapidly, and not a moment too soon.

As AI takes over more of the analytical and procedural work that used to require human time and attention, the skills that remain distinctly human are precisely the ones that were previously undervalued. The ability to read a room. To know what someone needs to hear versus what they want to hear. To navigate conflict without destroying a relationship. To make an ethical judgment call when the data points in one direction but your gut and your values point in another.

These are not backup skills for people who couldn't compete technically. They are the primary skills for the next era of work. And the people who have been developing them, sometimes without even realizing it, will find themselves in higher demand as AI handles more of the technical layer beneath them.

> ***Soft skills are no longer a nice addition to your resume. They are the primary currency of the AI era.***

WHAT THIS MEANS FOR YOU, RIGHT NOW

None of this requires a career overhaul tomorrow. What it requires is an honest assessment and a shift in how you think about your own development going forward.

The winners in the AI era are not a special class of people born with some advantage you missed out on. They are people who made a decision, often quietly, often without fanfare, to keep moving when it would have been easier to stay still. They read one more book. They tried one more tool. They asked one more question. They stayed curious when comfort was available.

That decision is available to you right now. Today. And the cost of making it is far smaller than the cost of waiting to see what happens.

TRY THIS RIGHT NOW

Personal Reflection

When was the last time I learned something completely new, not because my job required it, but because I was genuinely curious? What does that tell me about where I am right now?

AI Conversation Starter

I work in [your industry] as a [your role]. Based on what you know about the future of this field, what are the three human skills I should be developing most deliberately right now, and give me a specific 30-day plan to start?

Family Conversation

Ask your child tonight: what is one thing you taught yourself this year that no teacher assigned you? Then ask them how they figured it out.

DO THIS NOW

1. Write down three things you know how to do today that you didn't know five years ago. Then ask yourself: what is on your learning list for the next five years? If you don't have an answer, that is the problem to solve first.
2. Find one person in your life who is genuinely thriving amid change right now and ask them one specific question: what is the most important mindset shift you made to get here?
3. Replace 30 minutes of passive screen time this week with something that makes you a beginner again: a new podcast, a short course, a skill you've always meant to try. Notice how it feels to not already know the answer.

NEXT: Chapter 3 — You Don't Need to Code. You Need to Think.

PART TWO

The Adult Playbook

CHAPTER 3

You Don't Need to Code. You Need to Think.

How prompt literacy and clear thinking are the new power skills, for everyone

---

A colleague of mine spent three months convinced that AI had nothing to offer her. She was a gifted writer, a trained communicator, and she had built her career on the quality of her words. The idea that a machine could do what she did felt insulting. So she ignored it. Until a competitor started producing twice the output at half the cost using the exact tools she had dismissed. That got her attention.

Within six weeks of actually using the tools, she told me something I have heard dozens of times since: I wish I had started sooner. Not because AI replaced her writing. It didn't. Because it freed her to do more of the thinking and less of the grinding, and the quality of her work went up, not down.

That story repeats across every industry, every role, and every age group. The people who engage with AI tools early and intentionally come out ahead. The people who wait, whether out of fear, pride, or simply inertia, hand that advantage to someone else.

> ***You don't need to understand how AI works. You need to understand how to work with it.***

THE BIGGEST MYTH IN THE AI CONVERSATION

The most damaging myth circulating right now is that the AI era belongs to tech people. That myth is keeping millions of capable people on the sidelines of the most important shift of their professional lives. The truth is almost the opposite. The people extracting the most value from AI tools right now are not primarily engineers and developers. They are communicators, strategists, educators, entrepreneurs, and healthcare workers who learned one fundamental skill: how to have a productive conversation with an AI system. That skill is called prompt literacy, and it is the most practical and immediately useful capability you can develop right now, regardless of your technical background.

WHAT PROMPT LITERACY ACTUALLY MEANS

Prompt literacy is the ability to communicate with an AI tool in a way that produces useful, accurate, and relevant results. A well-crafted prompt gives the AI four things: context about who you are and what you are working on, a clear description of what you need, any relevant constraints or preferences, and an indication of what a good result looks like. That is not a technical skill. It is a communication skill. And if you have ever written a good email, briefed a colleague, or explained a problem clearly to someone, you already have the foundation.

10x *productivity increase reported by professionals who use AI tools with strong prompt skills versus those using the same tools without guidance. Stanford HAI, 2024.4*

THE THINKING LAYER NOBODY TALKS ABOUT

Here is where the real competitive advantage lives. AI tools are extraordinarily good at generating, summarizing, organizing, and refining information. What they are not good at is knowing which information matters, whether the output makes sense in context, and what to do with the result. That layer: the judgment layer, the so-what layer, is entirely yours.

The person who uses AI to generate ten possible approaches to a problem and then applies their own expertise, experience, and ethical judgment to choose the right one is not being replaced by AI. They are being amplified by it. The output of an AI tool is only as good as the thinking you bring to evaluating it. Garbage in, garbage out is still true. But good thinking in, excellent output out is equally true.

> ***AI makes you faster. Thinking makes you better. The combination makes you unstoppable.***

STARTING RIGHT NOW, TODAY

The fastest way to develop prompt literacy is not to take a course or watch a tutorial series. It is to use the tools on real problems with real stakes. Start with something that actually matters to you. Give an AI tool the full context and see what it does. You will not get it perfect the first time. That is fine. The learning happens in the iteration. Each refinement teaches you something about how to communicate more effectively, and that skill transfers to every tool you will ever use because the fundamental skill is clarity of thought, not knowledge of any particular platform.

WHAT THIS MEANS FOR YOUR CHILDREN

If you are a parent reading this, here is the most important thing to understand about AI and your children: the question is not whether they will use these tools. They already are. The question is whether they are using them to avoid thinking or to extend it. A child who uses AI to write their essay without engaging their own mind is building a dangerous dependency. A child who uses AI as a research assistant and a first-draft generator that they then critically evaluate and improve is building exactly the right relationship with the most powerful tool of their generation. The habit of using AI well starts with the habit of thinking well. And that starts at home.

TRY THIS RIGHT NOW

Personal Reflection

What is one task I do regularly at work or at home that I have never tried to do with an AI tool? What has stopped me, and is that reason still valid?

AI Conversation Starter

I am going to give you a real problem I am working on right now: [describe your actual problem in detail]. Help me think through it, give me three different approaches, and then tell me what questions I should be asking that I have not thought of yet.

Family Conversation

Sit down with your child and pick one thing they are curious about right now. Use an AI tool together to explore it, then talk about what the AI got right, what it missed, and what you both still want to know.

DO THIS NOW

1. Choose one real task this week, something that actually matters, and run it through an AI tool with a detailed prompt. Compare the result to what you would have produced alone. Note what surprised you.

2. Practice writing one better prompt per day for the next two weeks. Give more context, be more specific, and ask for exactly what you need. Watch how the quality of responses changes.

3. Find one colleague, friend, or family member who has not yet engaged with AI tools and share one practical use case with them. Teaching it is the fastest way to solidify your own understanding.

NEXT: Chapter 4 — Your Career Audit. A Blunt Conversation.

CHAPTER 4

Your Career Audit. A Blunt Conversation.

How to honestly assess where you stand, and what to do about it before someone else decides for you

---

Nobody likes to think their job might be at risk. It is uncomfortable, it triggers anxiety, and it runs directly against the very human tendency to assume that things will more or less continue as they are. I understand that. I have felt that resistance myself. But the people who have navigated major career transitions successfully share one thing in common: they did the honest audit before they had to.

They did not wait for the restructuring announcement. They did not wait until the job postings in their field dried up. They sat down with themselves, asked the hard questions, and made deliberate moves while they still had options. That is the conversation this chapter is about. Not to scare you. To give you accurate information so you can move with intention.

> ***Most people who get blindsided by disruption weren't unaware it was coming. They were unwilling to take it seriously until it was too late.***

THE THREE QUESTIONS THAT TELL YOU EVERYTHING

A career audit starts with three honest questions. First: how much of what I do every day involves routine pattern recognition, data processing, or following a defined process? Second: how much of my value comes from human judgment, relationships, creativity, or contextual wisdom that cannot be easily replicated? Third: if my company could automate 30% of my role tomorrow at lower cost, would they? This is the question most people avoid because the answer is often yes. But knowing the answer gives you agency. It tells you where the floor is and how much time you have to build higher.

> **41%** *of employers plan to reduce their workforce in areas where AI can automate tasks by 2030. World Economic Forum, Future of Jobs Report, 2025.1*

THE VULNERABILITY SPECTRUM

Not all roles are equally at risk, and it is worth understanding where your work sits on the spectrum. At one end are roles that are highly routine, heavily process-driven, and primarily involve the manipulation of information according to defined rules. These roles are already being significantly disrupted. At the other end are roles that require constant contextual judgment, complex human interaction, ethical reasoning, physical presence, or creative synthesis across unpredictable domains. Most people sit somewhere in the middle, and that middle is where the most actionable territory lies. The strategic move is to understand which parts of your role are automatable and which are genuinely hard to replace, then lean into the latter.

YOUR INDUSTRY IS NOT YOUR DESTINY

One of the most liberating realizations you can have about the AI transition is that your industry is not your destiny. The skills that make you effective in your current role, the communication, judgment, problem-solving, relationship-building, and domain knowledge you have accumulated, are largely transferable. They travel with you. Your existing expertise is the foundation. The question is what you are building on top of it.

> ***Your skills travel with you. The question is not whether your industry is safe. It is whether you are building something that is.***

BUILDING YOUR ONE-PAGE CAREER AUDIT

Take one page and divide it into four sections. In the first section, write down the five things you spend the most time on in your current role. In the second, mark each one as high, medium, or low automation risk based on how much human judgment it requires. In the third, write down three to five things you bring to your work that are genuinely hard to replace. In the fourth, write down one skill you could start building in the next 90 days that would shift your profile toward the resilient end of the spectrum. One page. Four sections. Honest answers. That is a career audit. It requires thirty minutes and the willingness to be honest with yourself about where you actually stand.

TRY THIS RIGHT NOW

Personal Reflection

If I am completely honest with myself, which parts of my job make me genuinely irreplaceable right now, and which parts am I doing primarily out of habit or comfort rather than unique value?

AI Conversation Starter

Here is a description of my current role and the main things I do day to day: [describe your role]. Assess which parts are most and least vulnerable to AI automation over the next three to five years, then give me a specific 90-day plan to start building in the right direction.

Family Conversation

Talk with your partner, spouse, or a trusted friend tonight: if one of our jobs changed dramatically in the next two years, what would our plan be? Have the conversation before you need to have it.

DO THIS NOW

1. Complete your one-page career audit this week using the four-section framework from this chapter. Be honest. The point is accurate information, not comfort.
2. Identify one adjacent skill, certification, or area of knowledge that would meaningfully increase your resilience in an AI-integrated version of your field. Write it down and set a start date.
3. Set a calendar reminder 90 days from today to review your audit. Note what has changed in your industry, in your role, and in your own development. Adjust accordingly.

NEXT: Chapter 5 — Robot-Proof. The Skills That Actually Last.

CHAPTER 5

Robot-Proof. The Skills That Actually Last.

Why the most human things about you are your greatest competitive advantage

---

I once sat in a meeting where someone proudly announced that their company had replaced its entire customer feedback analysis team with an AI tool. Faster, cheaper, more consistent, they said. And they were right on all three counts. What the tool could not do, and what became painfully obvious six months later when customer satisfaction scores started sliding, was notice that the feedback had an emotional undertone the numbers were not capturing. It took a human being with enough emotional intelligence to read between the lines and enough relational courage to bring that finding to leadership to turn things around. That person became one of the most valued people in the organization. Not despite the AI tools around them. Because of the contrast they provided.

> ***AI can generate a performance review. It cannot sit with someone who is struggling and know exactly what they need to hear.***

THE FOUR CAPABILITIES THAT CANNOT BE AUTOMATED

Research from Stanford, Harvard, McKinsey, and the World Economic Forum consistently identifies the same cluster of human capabilities as the most resilient in the face of automation.5

Emotional intelligence: the ability to perceive, understand, manage, and apply emotional information in yourself and in your relationships with others. Complex communication: the ability to convey ideas that matter to people who do not already agree with you, across differences in background, belief, expertise, and perspective. Ethical judgment: the ability to make principled decisions when the data points one way and your values point another. Creative problem-solving: the ability to connect ideas across domains and generate novel solutions to problems that do not have textbook answers. These are not backup skills for people who couldn't compete technically. They are the primary skills for the next era of work.

Top 5 *skills employers seek most by 2030: analytical thinking, resilience, leadership, creativity, and AI literacy. World Economic Forum, Future of Jobs Report, 2025.1*

WHY THESE SKILLS ARE MORE VALUABLE NOW THAN EVER

Here is the counterintuitive truth about the AI era: the more AI absorbs routine analytical and procedural work, the more valuable distinctly human capabilities become. This is not wishful thinking. It is basic economics. When the supply of something increases dramatically, the relative value of what it cannot replace goes up. AI is flooding the market with competent output. The scarcity is no longer in the ability to produce. The scarcity is in the ability to know which things are worth producing, to apply the judgment that makes competent output excellent, and to build the trust that makes people choose you over the machine.

> ***The goal is not to compete with AI. It is to become the person who uses AI better than anyone else while doing the things AI cannot do at all.***

DEVELOPING THESE SKILLS IS NOT PASSIVE

Emotional intelligence grows when you seek feedback on how your behavior lands on other people and you take that feedback seriously. Communication improves when you write more, speak more, and get honest feedback on what is landing. Ethical judgment sharpens when you engage with genuinely hard moral questions and build the habit of asking not just can we do this but should we. Creative problem-solving expands when you read outside your field, give yourself problems without obvious solutions, and expose yourself to diverse ideas and perspectives. None of this happens by accident. All of it is available to you, starting today, with no special equipment or credentials required.

THE COMBINATION IS THE ADVANTAGE

The most powerful professional profile in the AI era is not the person who refuses to use AI tools. Nor is it the person who uses AI for everything and has let their human capabilities atrophy. It is the person who combines strong AI literacy with strong human capabilities, who can move fluidly between the analytical power of the tools and the judgment, creativity, and relational depth that no tool can replicate. That combination is rare right now. The window to develop it deliberately is open. It will not stay open forever.

TRY THIS RIGHT NOW

Personal Reflection

In the last month, when did I do something that required genuine human judgment that a machine could not have replicated? What does that tell me about where my real value actually lives?

AI Conversation Starter

Give me a 30-day practice plan to develop stronger emotional intelligence and communication skills specifically for someone in my role as [your role]. Make it practical, daily, and realistic for someone with a full schedule.

Family Conversation

Ask your kids this week: what is something a robot could never do as well as a real person? Then ask them why. Listen carefully to the reasoning, not just the answer.

DO THIS NOW

1. This week, have one conversation you have been avoiding. Not because you enjoy conflict, but because practicing the discomfort of real human communication is how emotional intelligence grows. Notice what happens.

2. Find one ethical dilemma that AI is currently creating in your industry and form your own informed opinion on it. Not what your company thinks. Not what the media says. What do you actually believe, and why?

3. Read one book this quarter specifically about human psychology, leadership, communication, or creativity. The human layer is what needs developing most, and books are still one of the best tools for that.

NEXT: Chapter 6 — The School System Wasn't Built for This World.

PART THREE

The Parent Playbook

CHAPTER 6

The School System Wasn't Built for This World.

An honest conversation about what traditional education was designed for, and what it is missing now

---

I want to be careful here, because this chapter is easy to misread. This is not an attack on teachers. Most teachers are doing genuinely heroic work inside a system that was not designed for the world their students are about to enter. They are working with outdated tools, overcrowded classrooms, standardized tests that measure the wrong things, and curricula that lag years behind what the market and the culture actually need. The failure is not theirs. The failure is structural. And structural failures require structural honesty.

> ***The school system was designed to produce reliable employees for a stable industrial economy. That economy no longer exists.***

WHAT THE SYSTEM WAS BUILT FOR

The architecture of the modern school system was largely established during the late nineteenth and early twentieth centuries, a period defined by the needs of the industrial economy. Factories needed reliable, punctual, literate workers who could follow instructions, perform defined tasks, and function predictably within hierarchical organizations. Schools were designed, explicitly and intentionally, to produce those workers. The model worked for that world. The economy that exists today, and even more so the economy your child will enter, looks almost nothing like the one that system was built for.

10-15 Years *typical lag time between major societal shifts and meaningful curriculum change in public education systems. OECD Education at a Glance, 2024.6*

THE CURRICULUM LAG IS REAL

Educational systems are among the slowest-moving institutions in any society. When the shelf life of specific technical knowledge is measured in years rather than decades, a fifteen-year curriculum lag is not a minor inconvenience. It is a generational liability. A 2024 survey by the National Education Association found that fewer than 20% of K-12 teachers reported feeling adequately prepared to teach AI literacy or technology ethics.7 Not because they lack the intelligence or the dedication. Because the training, the tools, and the curriculum support simply have not arrived yet.

WHAT THE SYSTEM DOES WELL

In fairness, traditional education does several things well that deserve acknowledgment. It provides structure, socialization, and the development of foundational literacies that remain essential: reading, writing, mathematics, and basic scientific reasoning. It exposes young people to a breadth of subjects that helps them discover interests and aptitudes they might not have found on their own. It provides the credentialing that still opens doors in higher education and many professional fields. And at its best, it provides mentorship, community, and the experience of learning alongside others that shapes character in ways no online course can fully replicate.

> ***Your child's teachers are doing their best inside a system that has not caught up. That is not a criticism. It is your cue.***

WHERE THE GAP FALLS TO YOU

Here is the practical reality for every parent reading this: the gap between what the school system is currently teaching and what the next decade will actually require is real, it is significant, and for the foreseeable future it falls to families to fill it. That is not an unfair burden. In many ways it is simply a return to what families have always known: that parents are a child's first and most important educators. The question is not whether to supplement your child's education. It is how to do it deliberately, practically, and without turning your home into a second school that creates more pressure than it relieves.

TRY THIS RIGHT NOW

Personal Reflection

What skills am I hoping my child develops in school that I have never actually had a direct conversation with their teacher about? What would it look like to have that conversation this semester?

AI Conversation Starter

My child is [age] years old and attends [type of school]. Based on what you know about current curriculum gaps and the skills needed for the AI era, what are the three most important things their school is likely not teaching them, and how can I help fill those gaps at home?

Family Conversation

Ask your child this week: what do you wish you could learn in school that they don't teach? Then actually listen. The answer will tell you something important.

DO THIS NOW

1. Schedule a meeting with your child's teacher this semester, not about grades, but about future skills. Ask one direct question: what is the school doing to prepare students for an AI-integrated world? The answer, whatever it is, will be useful.

2. Research one supplemental education option that builds future-ready skills: a coding camp, a project-based learning program, a debate club, an entrepreneurship course, or a creative arts program that develops problem-solving.

3. Start a weekly family learning conversation. Pick one topic that none of you know much about, spend ten minutes researching it together using any tools available, and talk about what you found.

NEXT: Chapter 7 — What to Stop Optimizing Your Kid For.

CHAPTER 7

What to Stop Optimizing Your Kid For.

The metrics that used to matter, why they are no longer enough, and what to measure instead

---

My daughter once came home with a test score that made her cry. Not a failing grade. A B-plus. She was devastated. And I had to sit with a complicated mixture of feelings: pride in how much she cared, concern about the pressure she was putting on herself, and a deeper discomfort I could not quite name at the time. It took me a while to realize what it was. I had been part of building the belief system that caused her distress. By celebrating grades, by making academic performance the primary visible measure of her effort and value, I had taught her to optimize for the wrong thing.

> ***A child who gets straight A's by following every rule perfectly may be less prepared for the AI era than a child who gets B's but never stops asking why.***

WHAT GRADES WERE DESIGNED TO MEASURE

Grades measure a specific and relatively narrow set of capabilities: the ability to absorb defined content, reproduce it accurately under time pressure, and demonstrate compliance with the structure of the task. Grades are not useless. Academic performance still matters for college access. The problem is when grades become the primary or exclusive metric by which a child understands their own worth and capability. Because what grades do not measure is often more important than what they do. They do not measure curiosity, resilience, the ability to ask a great question, or the ability to work through a problem that has no right answer. And those are precisely the capabilities the AI era rewards most.

94% *of employers prioritize critical thinking, communication, and adaptability over technical credentials in long-term hiring decisions. LinkedIn Workforce Report, 2024.*[8]

THE METRICS THAT ACTUALLY MATTER

Five indicators are far more predictive of long-term success in the AI era than a GPA: curiosity, whether your child asks questions that go beyond what is required; resilience, how your child responds to failure and frustration; initiative, whether your child takes ownership of their own learning; communication, whether your child can explain what they know to someone who does not already know it; and ethical reasoning, whether your child thinks about the why and the so what, not just the what. These five indicators will serve your child long after any grade is forgotten.

> ***The most future-ready children are not the most pressured. They are the most curious, the most resilient, and the most willing to be a beginner again.***

RAISING A CHILD WHO THINKS, NOT JUST PERFORMS

The shift from optimizing for performance to developing for capability is not about lowering standards. It is about raising the right ones. It means celebrating the question as much as the answer. It means making failure a data point rather than a verdict. It means creating space for your child to pursue something they care about deeply, even if it is not directly connected to their school curriculum, because passion-driven learning builds exactly the intrinsic motivation that institutional learning often struggles to sustain.

A WORD ABOUT PRESSURE

One of the most damaging things we can do to a child in the AI era is layer our own anxiety about the future onto their development. The most future-ready children are not the most pressured ones. They are the most curious ones. The ones who know how to be a beginner, who are not destroyed by imperfection, and who have a strong enough sense of who they are to navigate ambiguity without losing themselves. Building that in your child starts with believing it yourself.

TRY THIS RIGHT NOW

Personal Reflection

When my child brings home a disappointing grade or result, what is my first response? And does that response build the capabilities I actually want them to have, or does it reinforce the metrics I am trying to move away from?

AI Conversation Starter

My child is [age] and their strengths seem to be [describe strengths]. What are the most important habits and capabilities I should be nurturing right now for long-term success, and what are some specific ways to build them at home without adding more pressure?

Family Conversation

Tell your child about a time you failed at something important and what you learned from it. Make it real, not sanitized. Let them see that failure is survivable and instructive.

DO THIS NOW

1. This week, find one moment to praise your child specifically for their curiosity, their effort, or their resilience rather than their result. Notice how they respond differently to being seen for how they approached something versus what they produced.

2. Introduce one open-ended, impossible-seeming problem at the dinner table this week and resist the urge to give them the answer. Let them wrestle with it. The discomfort of not knowing is part of the learning.

3. Look at your child's weekly schedule and ask honestly: how much of it is building real skills and character versus filling a resume? If the balance is off, pick one thing to add that genuinely develops them and one thing to remove that is just noise.

NEXT: Chapter 8 — What to Start Doing At Home Right Now.

CHAPTER 8

What to Start Doing At Home Right Now.

Practical, age-specific actions that build the capabilities school is not yet teaching

---

The most future-proof thing you can build in your home is not a faster internet connection or a better laptop. It is a culture of curiosity. A household where learning is celebrated, failure is survivable, questions are welcomed, and the adults model the exact behaviors they want to see in their children.

> ***You are not trying to turn your home into a second school. You are trying to make your home a place where the right kind of learning happens naturally.***

FOR YOUNGER CHILDREN: AGES 5 TO 10

At this age, the most important thing you can build is the love of learning itself. Read together as an adventure, not just a wind-down routine, and ask questions about the story: what do you think happens next, why do you think the character made that choice, what would you have done differently? Let them build things and let those things fail. The act of building something, having it not work, figuring out why, and trying again is one of the most powerful learning experiences available at any age. Resist the urge to fix it for them. Introduce AI tools in supervised, playful ways: ask a question together and look at what the AI says, then talk about what it got right, what it got wrong, and how you would check.

Ages 5-10 *the window where habits of curiosity, resilience, and love of learning are most powerfully shaped. Harvard Center on the Developing Child.9*

FOR MIDDLE SCHOOLERS: AGES 11 TO 14

Middle school is the highest-leverage window for parents. The habits built here tend to persist. Encourage a personal project: something they care about deeply that requires research, creativity, and iteration. It can be anything: building a website, writing a short film, designing a game, starting a small business. The content matters less than the process. Teach them to use AI as a research partner, not a homework machine. Make the distinction explicit and revisit it regularly: using AI to think is powerful, using AI instead of thinking is a trap. Have direct conversations about the future of work. Children who understand the why behind the development they are being asked to do are dramatically more likely to engage with it genuinely.

FOR HIGH SCHOOLERS: AGES 14 TO 18

Help them build AI literacy deliberately: not just using tools for homework, but understanding what they are good at, where they fail, and what ethical questions they raise. Introduce real-world financial and entrepreneurial thinking. The ability to understand money, create value, and think like someone who builds things rather than just consumes them is foundational for any path they choose. Help them develop a strong sense of who they are beyond their academic performance. What do they believe? What do they value? What kind of person do they want to be? These are not extracurricular questions. They are the foundation of the judgment, resilience, and ethical reasoning that will serve them for the next fifty years.

> ***The child who knows who they are and what they stand for will always have an edge over the child who only knows what they have achieved.***

FOR ALL AGES: THE FAMILY PRACTICES THAT MATTER MOST

Ask better questions at the dinner table. Not how was your day, but what is something you learned today that surprised you, or what is a problem you are trying to figure out? Model the behaviors you want to see. If you want your child to be curious, be curious in front of them. Create margin in their lives. The overscheduled child who moves from activity to activity with no unstructured time is not being developed. They are being managed. Creativity, reflection, and deep self-directed thinking require space. Protecting that space is one of the most countercultural and most important things you can do as a parent right now.

TRY THIS RIGHT NOW

Personal Reflection

Looking honestly at my child's weekly life, how much of it is building genuine capability versus performing for external metrics? What would I change if I were optimizing for who they become rather than how they look on paper?

AI Conversation Starter

My child is [age] and their biggest strengths are [strengths] and their biggest areas for growth are [areas]. Design a realistic 60-day home learning plan that builds future-ready skills without adding overwhelming pressure to their existing schedule.

Family Conversation

Pick one problem in your community this week and spend twenty minutes as a family brainstorming how you would solve it. No right answer. No grades. Just thinking together.

DO THIS NOW

1. Start a question jar in your kitchen this week. Everyone in the family drops in one question they are genuinely curious about. Once a week, pull one out and explore it together. Make it a ritual, not a project.
2. Let your child use an AI tool this week to plan something they care about: a party, a trip, a creative project, a business idea. Then review the output together and talk about what the AI got right, what it missed, and what they would change.
3. Remove one structured activity from your child's schedule this month and protect that time as unstructured. Notice what they do with it. The answer will tell you something important about who they are becoming.

NEXT: Chapter 9 — The Church Can't Afford to Sit This One Out.

PART FOUR

The Moral Frontier

CHAPTER 9

The Church Can't Afford to Sit This One Out.

Why faith communities are uniquely positioned to lead the most important conversation of our time

---

Every significant moral crisis in modern history has had a moment when the faith community had to decide: engage or retreat. During the abolition movement, it was the churches that provided both the moral framework and the community infrastructure that made the movement possible. During the civil rights era, faith leaders stood at the center of the most consequential social transformation in American history. In each case, the communities that showed up with wisdom, compassion, and a values-first framework earned something money cannot buy: the deep trust of the people who needed it most. AI is the next great moral frontier. And the church has a choice to make right now.

> ***The church doesn't need a new message to engage AI ethics. It needs to apply the timeless one it already has to a new and urgent context.***

WHAT HAPPENS WHEN ETHICS ARE LEFT TO TECH COMPANIES ALONE

Algorithmic bias is producing discriminatory outcomes in hiring, lending, healthcare, and criminal justice. AI-generated deepfakes are being used to destroy reputations and spread disinformation in ways that are difficult to detect and nearly impossible to reverse.

Autonomous systems are making consequential decisions about people's lives without meaningful human accountability or recourse. These are not hypothetical concerns. They are current events. And they are accelerating. When the people building the most powerful technology in human history are primarily accountable to shareholders and market forces, a moral counterweight is not optional. It is essential. Faith communities, with their deep roots, their trusted relationships, and their frameworks built on human dignity, truth, justice, and accountability, are uniquely positioned to provide it.

> **72%** *of Americans trust their faith community as a source of guidance on ethical questions more than the government, media, or technology companies. Pew Research Center, 2024.10*

THE THEOLOGICAL CASE, WITHOUT THE LECTURE

Every major faith tradition has a framework that applies directly to how AI should and should not be used, and none of it requires new theology. The concept of human dignity speaks directly to the risk of AI systems that treat people as data points rather than human beings. The concept of stewardship speaks directly to the ethical obligations of those building and deploying AI systems. The concept of truth speaks directly to the deepfake crisis and the disinformation ecosystem. The concept of justice speaks directly to algorithmic bias and the risk that AI systems encode and amplify existing inequalities. The message is already there. It has been there for centuries. What is needed is the courage and the clarity to apply it.

WHAT CHURCHES CAN ACTUALLY DO RIGHT NOW

Host an AI literacy night for your congregation. Create a safe space for the fear, the confusion, and the genuine concern that most people are carrying about this shift but have nowhere to put. Create conversation frameworks for small groups: the questions AI raises about work, identity, truth, and what it means to be human are exactly the questions faith communities are built to explore together. Partner with the parents and educators in your congregation who are already doing this work. Equip your leaders to speak with intelligence and compassion on AI ethics. Pastors do not need to become technologists. But they do need to be able to engage with the questions their congregants are carrying.

> ***The congregation that shows up with wisdom rather than condemnation will earn the trust of the generation that needs it most.***

THE GENERATION THAT NEEDS THIS MOST

Generation Z and Generation Alpha are navigating the most complex information environment in human history with almost no ethical guidance. They are using AI tools daily. They are making decisions about their identity, their beliefs, and their future in a context where the line between real and synthetic, between authentic and manufactured, is increasingly blurred. The faith community that shows up to that generation with wisdom rather than condemnation, with curiosity rather than fear, and with a values framework that gives them something solid to stand on will earn an influence and a trust that no marketing budget can manufacture. The window to lead rather than react is open right now. It will not stay open indefinitely.

TRY THIS RIGHT NOW

Personal Reflection

What would it look like for my faith community to lead on AI ethics rather than react to it? What is one specific conversation I could initiate in my church, small group, or community of faith this month?

AI Conversation Starter

From the perspective of [your faith tradition], what ethical principles should guide the development and use of artificial intelligence? Give me a framework that a pastor, elder, or community leader could actually use in a sermon, small group, or community conversation.

Family Conversation

Ask your family: what values should guide how humans build and use AI? Where do those values come from, and how do we make sure they are actually followed? Let the conversation go wherever it goes.

DO THIS NOW

1. Bring up the topic of AI ethics in your next small group, leadership meeting, or elder board conversation. Even informally. The goal is not to have all the answers. The goal is to open the conversation before someone else defines it for your community.

2. Research one current AI ethics issue that is already affecting people in your congregation: job displacement, deepfakes, algorithmic bias, or data privacy. Come prepared to explain it clearly and compassionately.

3. Propose one community initiative to your church leadership: an AI literacy night, a parent workshop on preparing children for the AI era, or a youth conversation on digital ethics and what it means to live with integrity online. Start small. Start soon.

NEXT: Chapter 10 — The Only Real Risk Is Standing Still.

PART FIVE

The Decision

CHAPTER 10

The Only Real Risk Is Standing Still.

A final challenge, a clear decision, and the one thing that separates the people who thrive from the ones who don't

---

Here is something true about every major technological revolution in history: the people who look back on it with regret are almost never the ones who moved too early. They are the ones who waited too long. The ones who wanted more certainty before they acted, who needed to see how things played out before they committed, who told themselves they would get started when things settled down a little. Things did not settle down. They never do.

You have now read nine chapters. You know more about the AI transition than the vast majority of people who are navigating it right now. That knowledge is worth exactly nothing if it stays in your head. The entire point of understanding a wave is to know how to ride it.

> ***The people who get left behind in every major technological shift aren't the least capable. They're the least willing to move.***

WHAT EVERY REVOLUTION HAS TAUGHT US

The printing press did not destroy knowledge. It democratized it, and the people who learned to use it earliest shaped the intellectual landscape of the next two centuries. The Industrial Revolution did not eliminate work. It transformed it. The internet did not destroy commerce, communication, or community. It remade all three. AI will follow the same arc. It will displace work that depends on routine and repetition. It will create work that requires judgment, creativity, and human connection. And it will reward, as every previous revolution has rewarded, the people who moved before the movement became obvious.

3-5 Years *the window most economists and workforce researchers identify as critical for individuals to adapt before AI transitions become irreversible in most industries. McKinsey Global Institute, 2025.11*

THE DECISION IN FRONT OF YOU

Every person who reads this book will make one of three choices. The first is to do nothing: to find the content interesting, perhaps even urgent-feeling, and then return to the routines and the comfort of the familiar and let the urgency fade. Most people will make this choice. Not because they are incapable. Because inertia is powerful, and there is always a reason to wait just a little longer.

The second choice is to make a small start: to take one action, develop one skill, have one conversation that was not happening before. Small starts are underrated. They build momentum, they create evidence of your own capability, and they open doors that were not visible from the shore.

The third choice is to commit: to treat the shift you now understand as the real and urgent challenge it is, to build it into your schedule and your priorities, to share it with the people around you, and to decide that the future you want for yourself and your family is worth the discomfort of moving before you feel fully ready.

All three choices are available to you right now. Only one leads to the future you want.

THE HOUSE IS NO LONGER JUST ON FIRE

At the beginning of this book, I told you the house was on fire. Now, at the end, I want to change the image slightly. Because the fire is real, but so is what is being built on the other side of it. Every major transition in human history has been followed by a period of extraordinary creation. New industries, new opportunities, new ways of working and living and connecting that were not possible before the disruption cleared the ground.

The people who are positioned to build in that space are the ones who moved through the fire rather than away from it. The ones who developed the skills, built the habits, equipped their children, engaged their communities, and arrived at the other side still standing and still growing. That is the opportunity in front of you.

> ***The future belongs to the people who build it. Not the ones who waited to see how it turned out.***

ONE LAST THING

Share this book. Not just as a recommendation. As a conversation starter. The friend who is anxious about their job security. The parent who is worried about their child's future. The pastor who knows something important is happening but does not quite have the language for it yet. The colleague who dismisses AI as hype because engaging with it feels overwhelming. The most powerful thing you can do with what you have read is not to apply it privately and keep it to yourself. It is to bring other people into the conversation. The AI transition is not a problem any individual navigates alone. The communities, families, workplaces, and congregations that navigate it together will come out of it far stronger than the ones where everyone figured it out in isolation.

Start the conversation. Today. Before the urgency fades and the familiar reclaims its grip.

The house is on fire. You know where the exits are. And you know, now, what to build on the other side.

TRY THIS RIGHT NOW

Personal Reflection

What is the single most important thing I have read in this book that I have not yet acted on? What has stopped me? And is that reason more important than the outcome I say I want?

AI Conversation Starter

Based on everything in this book about AI, my career, my family, and my community, give me a specific 90-day action plan with weekly milestones. Make it realistic, make it specific, and make the first step something I can do today.

Family Conversation

Sit down with your family tonight and ask: what is one thing we are going to commit to doing differently, starting this week? Write it down. Put it somewhere visible. Check in on it in 30 days.

DO THIS NOW

1. Write down your single most important commitment from this book. One sentence. On paper. With a date. Then put it somewhere you will see it every day.
2. Share this book with one person in your life who needs it. Not as a gift, as a conversation starter. Call them, text them, or hand it to them directly and tell them why you thought of them.

3. Set a 90-day review date in your calendar right now. Three months from today, come back to the commitments you made and measure honestly how you have done. Then decide what comes next.

REFERENCES

All statistics and research findings cited in this book were accurate at time of publication. Readers are encouraged to visit source websites for the most current data.

1.. World Economic Forum. (2025). Future of Jobs Report 2025. World Economic Forum, Geneva, Switzerland. Retrieved from www.weforum.org/reports/future-of-jobs-report-2025. Cited for: 39% skills obsolescence; 92 million jobs displaced and 170 million new roles created; net gain of 78 million jobs by 2030; 41% of employers planning workforce reductions; top 5 skills by 2030; 59 out of 100 workers needing reskilling.

2.. London School of Economics and Political Science. (2023). Intellectual Curiosity and Career Resilience in Automated Industries. LSE Research Online. London, UK. Retrieved from www.lse.ac.uk/research. Cited for: curiosity as a top predictor of career resilience across industries facing automation.

3.. AARP Public Policy Institute. (2022). Staying Ahead: Upskilling Workers Over 50 in a Technology-Driven Economy. AARP, Washington, D.C. Retrieved from www.aarp.org/ppi. Cited for: workers over 50 who engaged in regular upskilling keeping pace with and outperforming younger colleagues in AI-adjacent roles.

4.. Stanford University Human-Centered Artificial Intelligence (HAI). (2024). AI Productivity and Prompt Skill Research Report. Stanford HAI, Stanford, CA. Retrieved from www.hai.stanford.edu. Cited for: 10x productivity increase for professionals using AI tools with strong prompt skills.

5.. McKinsey Global Institute; World Economic Forum; Harvard Business Review. (2023-2025). Future of Work Research Series. Various publishers. Sources include: McKinsey & Company, www.mckinsey.com/mgi; World Economic Forum, www.weforum.org; Harvard Business Review, www.hbr.org. Cited for: the four human capabilities most resilient to automation: emotional intelligence, complex communication, ethical judgment, and creative problem-solving.

6.. Organisation for Economic Co-operation and Development (OECD). (2024). Education at a Glance 2024: OECD Indicators. OECD Publishing, Paris, France. DOI: https://doi.org/10.1787/eag-2024-en. Cited for: 10-15 year typical lag between major societal shifts and meaningful curriculum change in public education systems.

7.. National Education Association (NEA). (2024). Teaching in the Age of Artificial Intelligence: National Survey of K-12 Educators. NEA, Washington, D.C. Retrieved from www.nea.org/research. Cited for: fewer than 20% of K-12 teachers reporting feeling adequately prepared to teach AI literacy or technology ethics.

8.. LinkedIn Workforce Insights. (2024). LinkedIn Workforce Report 2024: Skills, Hiring, and the Future of Work. LinkedIn Corporation, Sunnyvale, CA. Retrieved from www.linkedin.com/pulse/workforce-report. Cited for: 94% of employers prioritizing critical thinking, communication, and adaptability over technical credentials in long-term hiring decisions.

9.. Harvard Center on the Developing Child. (2023). Key Concepts in Early Childhood Development: Critical Windows for Learning. Harvard University, Cambridge, MA. Retrieved from www.developingchild.harvard.edu. Cited for: ages 5-10 as a critical window for shaping habits of curiosity, resilience, and love of learning.

10.. Pew Research Center. (2024). Faith Communities and Public Trust: Americans' Views on Ethical Guidance in a Technology-Driven World. Pew Research Center, Washington, D.C. Retrieved from www.pewresearch.org. Cited for: 72% of Americans trusting their faith community more than government, media, or technology companies on ethical questions.

11.. McKinsey Global Institute. (2025). The Future of Work in America: People and Places, Today and Tomorrow. McKinsey & Company, New York, NY. Retrieved from www.mckinsey.com/mgi/research. Cited for: 3-5 year window identified as critical for individuals to adapt before AI transitions become irreversible in most industries.

FUTURE-PROOF

ISBN: 978-1-962942-11-9

Copyright © 2026 Chauntelle Ricks. All Rights Reserved.

ALSO BY CHAUNTELLE RICKS

Devotional Works in the Faith-Based Community

COMING SOON

Blended

How Teaching Your Child About AI Will Teach You Everything You Need to Know

The follow-up to Future-Proof goes deeper into the home, the classroom, and the community. Hands-on activities for every age. A blueprint for faith communities and neighborhoods to build AI learning hubs. An honest look at what the new workforce will actually look like, and how to raise children who are ready to build it.

MORE FROM CHAUNTELLE RICKS

Continue your AI journey with these resources:

Edge Effect AI Community on Skool

A growing community of professionals, parents, and faith leaders applying AI with clarity, confidence, and purpose.

www.skool.com/edge-effect-ai-8371

Courses, Templates, and Tools

Practical AI resources you can put to work immediately, wherever you are in your journey.

www.stan.store/edgeeffectai

Speaking, Training, and Bulk Orders

Contact Us:
Chauntelle Ricks
ccharisseai@gmail.com

The people who get left behind in every major technological shift aren't the least capable.
They're the least willing to move.

Made in the USA
Monee, IL
04 November 2026

67376116R00039